Mom

High thanks I owe you for a lifetime of care,
Your gentle teachings, Your heart so fair;
You're given your soul and held no above,
And because of you I know how to love.

Love always
Patty and Dan

THE *Little Book* OF

~

ROSES

~

A CHARMING AND INSPIRATIONAL
GUIDE TO GROWING
AND USING ROSES

Editor: Jo Finnis

Contributors: David Squire, Mary Lawrence p.20-21, 24-25, 26-35, Jo Finnis

Designer: Sally Strugnell

Original Design Concept: Peter Bridgewater, Nigel Duffield

Picture Researchers: Leora Kahn, Miriam Sharland

Director of Production: Gerald Hughes

Typesetting: Julie Smith

Photography: ART RESOURCE, New York: p.11, 44. THE BETTMANN ARCHIVE, New York: p.16, 17. THE BRIDGEMAN ART LIBRARY, London, with acknowledgements to: the British Library, London, for *Roman de la Rose*/Harley ms, p.8; the V. & A. Museum, London for *The York and Lancaster Rose*/Ehret, p.12, and *Miniature of a Young Man Against a Rose Tree*/Hilliard, p.45; Chelsea Physic Garden for *Rosa Setigera*/Parsons, p.13; St Martin, Colmar, Germany, for *Madonna of the Rose Bower*/Schöngauer, p.43; and Christie's, London, for *Roses*/Fantin-Latour, p.46-47; Private Collection for *Rosa Gallica Regalis*/Redouté, p.48; Louvre, Paris, for *Gabrielle with Rose*/Renoir, p.51. JOHN GLOVER: p.49, p.61. HARRY SMITH COLLECTION: p.52-53, 54-55, 56 (cut-out), 57. DAVID SQUIRE: p.53 (cut-out).

CLB 3129

ISBN 0-86283-972-6

THE *Little Book* OF
ROSES

SMITHBOOKS

Ancient Origins

The earliest record of a rose is thought to be of a Damascene rose, a natural hybrid between *Rosa gallica* and *Rosa phoenicea*, found in frescoes at Knossos, a ruined city on the island of Crete and at one time capital of the Minoan civilization, about 3000-1100 BC. Also depicted on frescoes, this time in Pompeii and subsequently mentioned by the Greek historian Herodotus (484-425 BC), later known as the 'father of history', is the Autumn Damask Rose.

A Damascene rose was taken by early Christians to Abyssinia and planted in the Province of Tigre, where it was known as the Holy Rose (*Rosa sancta*). It spread and forms were introduced into Europe by returning Crusaders. Indeed, the famous French poem *Le Roman de la Rose* (The Romance of the Rose), an elaborate allegory on love and secular life begun in the latter half of the 13th century, later translated and completed by Geoffrey Chaucer, tells of roses being introduced into France from the lands of the Saracens.

The French poet and songwriter, Thibaut (sometimes Theobold) IV, Count of Champagne and Brie, and King of Navarre, returned to Provins in 1240 after two years in Palestine, taking roses with him. From the 13th to the 19th century, Provins was famous for medicinal roses. Many of these roses became known as Gallicas because of their assumed derivation in France, but were really forms of the Damask Rose.

King Midas, the legendary King of Phrygia, a former kingdom in western and central Asia Minor, is said to have grown the Autumn Damask Rose and introduced it into gardens in a district of Macedonia, where much later a form of it was used to produce rosewater and attar of roses in the Kazanlik region of Bulgaria.

Left: Le Roman de la Rose, *an allegorical poem on love and secular life, is depicted here in about 1500, the lover attaining the rose.*

9

Fair Mistresses

In the middle of the 12th century, the fair Rosamond was the mistress of Henry II of England. Legend tells that Henry's wife, Queen Eleanor of Aquitaine, on the pretence of offering silken thread poisoned her. The 16th-century poet, Samuel Daniel, wrote a poem on the sorrows of Rosamond called *The Complaynt of Rosamond*, and it is said that Rosa Mundi *(Rosa gallica versicolor)*, with crimson flowers striped white, was named after her.

Madame de Pompadour, mistress of King Louis XV of France, was frequently painted holding roses, especially the fragrant Belle de Crécy.

The Empress Josephine, wife of Napoléon I, acquired the chateau of Malmaison, Hauts-de-Seine, near Paris, where, among other plants, she assembled a vast collection of roses.

Left: *Madame de Pompadour, mistress of Louis XV of France, was frequently painted with roses - in this case by François Boucher.*

Rosa damascena 'Versicolor', the York and Lancaster Rose, is thought to be the rose around which the Dukes of York and Lancaster quarreled in the Temple Gardens.

*E*mblems

Perhaps the best known rose emblems are those of the houses of York and Lancaster during the Wars of the Roses in England. In 1455, a quarrel between the Dukes of York and Lancaster in the rose garden of the Temple in London initiated the war. The protagonists are said to have each plucked a rose as their emblem: the Yorkists selected a white rose, the Lancastrians a red one, but long before the war, both these families were associated with roses. On the Yorkists' side, a white rose was the emblem of Eleanor of Provence who, in 1235, married Henry III, her emblem descending to her son, Edward I. For the Lancastrians, the red rose was gained by Edmund, the second son of Henry III through his marriage in 1275 to Blanche, widow of Henry I of France.

After about 30 years, the brutal dispute was settled by a marriage between Henry Tudor, obscurely related to both sides, and Elizabeth of York, daughter of Edward. The Tudor Rose was used by Henry's son, Henry VIII, as an emblem.

Rosa setigera, the Prairie Rose, a sprawling and rambling shrub native to eastern and central North America, is the emblem of the American state of North Dakota.

Roses have been widely used as emblems in North America - indeed, they have been adopted by several states. The White Cherokee Rose *(Rosa laevigata)* is claimed by Georgia. Curiously, it is not native to North America, originating in China but naturalized in the southern states as early as 1780. In legend, it is associated with an Indian girl who was magically turned into a flower when captured by a hostile tribe. The district of Columbia has adopted the American Beauty Rose, while the Wild Prairie Rose (*Rosa setigera*) is claimed by North Dakota. However, *Rosa arkansana* (also known as *Rosa pratincola* or *Rosa suffulta)*, a North American native rose and at one time a persistent weed of prairie wheat fields, is also claimed by North Dakota. The Pasture Rose (*Rosa carolina)* is the state flower of Iowa.

A legend surrounds the blood-red petaled Grant Rose, which is supposed to have sprung from the blood of Mrs Grant, an early settler in Florida killed by a Seminole (native American Indian).

13

The Meanings of Flowers

The passing of messages through flowers was known in Turkey in the 1600s, but did not spread to Europe until 1716 when the celebrated letter writer and society poet, Lady Mary Wortley Montague, accompanied her husband to the Turkish Court in Constantinople. She learned how messages of love could be passed without recourse to letter writing or talking.

Returning to England in 1718, she immediately told friends of the meanings of flowers, but after quarreling with the poet and satirist, Alexander Pope, went to live abroad. The French were enthused by the romanticism of flowers and quickly took to using flowers to send quite complex messages, returning the idea to Britain during the reign of Queen Victoria through a book by Madame de la Tour called *Le Langage des Fleurs*. Some of its sentiments and messages were too lusty for the Victorians and needed to be tempered.

The Language of Roses

Burgundy Rose - *simplicity and beauty;*
unconscious beauty
Carolina Rose - *love is dangerous*
China Rose - *grace or beauty ever fresh;*
beauty is always new
Damask Rose - *brilliant complexion;*
beauty ever new
Dog Rose - *pleasure mixed with pain; simplicity*
Eglantine - *poetry; I wound to heal*
Moss Rose - *voluptuous love; confessions of love*
Multiflora Rose - *grace*
Musk Rose - *capricious beauty*
Provence Rose - *my heart is in flames*
Rose - *love and beauty*
White rose - *simplicity*
White and red roses together - *unity;*
warmth of the heart
White rosebud - *too young to love; girlhood;*
a heart ignorant of love
York and Lancaster Rose - *war*

15

Below: *A Victorian Valentine depicting a young lady deep in amorous thoughts, framed by voluptuous apricot roses.*

Valentines

Roses have long been featured in rhymes associated with St Valentine's Day, a fusion of pagan and Christian customs. The Roman feast of Lupercalia, celebrated on February 15, was a mating ritual when girls of a marriageable age had their names put into an urn for the local lads to select.

Little is known about St Valentine, other than there were at least two martyrs of that name, both dying for their religion on February 14. During the third century, St Julius I, Pope from 337-352, allotted a saint day to St Valentine, and when christianity reached Britain the Feast of Lupercalia was moved back one day and merged with St Valentine's Day. At one time, it was commonly believed that all birds chose their mates on this day.

This fertility festival gained popularity and gifts were given by men to women of their choice. It also became a custom to recite rhymes:

Roses are red and violets are blue,
Carnations are sweet and so are you.
Thou art my love, and I art thine,
I draw thee for my Valentine.

A variation was:

The rose is red, the violet blue, Gillies are sweet,
and so are you;
These are the words you bade me say,
For a pair of new gloves on Valentine's Day.

The gift of a pair of gloves was said to symbolize lasting love. Sending cards anonymously became popular, perhaps a legacy of the uncertainty and anticipation when a young man selected the name of a girl out of a Roman urn.

Right: *A lavish Valentine card dating from around 1910 which features the rose as the emblem of fond love.*

17

Apricot Roses and Chocolate Leaves

To make the roses, flatten out the dried apricots with your fingers (they are often shriveled and twisted) to faciliate slicing in half widthways. You will need a good, sharp knife for the job. Sandwich the apricot halves between two sheets of waxed (greaseproof) paper and with a rolling pin, roll out the fruit pieces as thinly as you can, to a thickness of a fraction of an inch (or 1 mm). To make the center of the rose, roll up one of the apricot halves to form a tight bud with the cut side inside. This will help the bud to stay in place, since the cut side of the apricot is sticky. Take a second apricot piece and wrap halfway round the rose center, again with the cut side inside to adhere. Add a further two 'petals' in the same way, overlapping the outer petal by half in each case. Ease the top edges of the petals outwards with the fingers to give a natural effect. Flatten the bases of the roses if necessary and chill for at least 20 minutes before use.

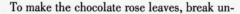

To make the chocolate rose leaves, break un-

sweetened or semisweet (plain or milk) chocolate into pieces and place in a bowl over a hot water bath to melt. Wipe leaves clean and, using a small paintbrush, paint the top side of the leaves with the melted chocolate. Lay over a rolling pin until set, then carefully peel away leaves.

Rose Turkish Delight

2½ cups (1¼ lb or 575 g) white cane sugar
⅔ cup (5 fl oz or 150 ml) triple-strength rosewater
1 oz or 25 g powdered gelatin
¾ cup (4 oz or 100 g) confectioner's (icing) sugar
½ cup (2 oz or 50 g) cornstarch (cornflour)

Put the sugar and rosewater in a deep, heavy pan. Heat gently, stirring occasionally, until the sugar is dissolved, then bring to the boil. Raise the temperature to 240°F (115°C), testing with a candy thermometer, and maintain for three minutes.

Meanwhile, dissolve the gelatin in 3½ table-spoons (50 ml) of water in a small saucepan, then bring to the boil. Add two drops of red food coloring to the sugar, if desired. Very carefully and slowly, add the boiling gelatin solution to the boiling sugar and incorporate using a balloon whisk. Boil for two minutes and set aside to cool

below 200°F (100°C). Lightly butter a 6 in (15 cm) square shallow pan. Pour the mixture into the pan and leave for a day in a cool place to set.

Mix together the confectioner's (icing) sugar and cornstarch (cornflour) and spread on a sheet of waxed (greaseproof) paper. Dip the base of the pan of Turkish Delight in hot water for a few seconds, then turn out onto the sugar. Spoon the sugar over the block. Cut the block into squares using a sharp knife coated in the sugar, frequently adding more sugar down the line of the knife cut. Ensure that each cube is well coated, then store for two days in a cool place to give a crisp texture to the coating.

Make another batch using water in place of the rosewater and adding a handful of coarsely chopped pistachio nuts, 3 drops of green food coloring and 6 drops of pistachio flavoring.

Rose and Pistachio Ice-cream

2 cups scented rose petals
1½ cups (10 fl oz or 300 ml) light (single) cream
½ cup (2 oz or 50 g) sifted confectioner's
(icing) sugar
2 tbsp (35 ml) rosewater
½ cup (2 oz or 50 g) pistachio nuts, shelled and
coarsely chopped

Remove the yellow bases of the petals. Place them in a pan together with the cream and sugar and bring slowly to the boil. Strain the cream mixture, taking care to extract as much liquid as possible from the petals. Incorporate the rosewater and leave to cool. Pour into a freezer-proof container, cover and freeze for at least an hour, together with an empty bowl. Beat the mixture in the bowl, return to the container and freeze for a further three hours. Transfer to the refrigerator for 30 minutes before serving.

Rose Fondant Shapes

⅔ cup (5 fl oz or 150 ml) triple-strength rosewater
2 cups (1 lb or 450 g) loaf or granulated cane sugar
2 tbsp (1 oz or 25 g) glucose
1 tbsp (15 ml) heavy (double) cream

Put the rosewater and sugar in a heavy pan and heat gently, stirring occasionally, until the sugar has dissolved. Bring the syrup to the boil, add the glucose and raise the temperature to 240/245°F (115°C), testing with a candy thermometer. Hold

24

this temperature for two minutes, then set aside to cool for 10 minutes. Lay out a marble slab, sprinkle with very little water and pour the syrup onto it. Using two spatulas, keep turning the syrup into the center until it changes its nature and becomes opaque and firm. Scrape what is now the fondant off the slab and set aside.

Clean the slab, dry thoroughly, then knead the fondant on its surface, adding a little heavy (double) cream on the end of a skewer. Leave the fondant to cool and become firm. Either press into a pliable candy mold or break into small pieces and hand mold into shapes. We made our shapes by rolling the fondant into a 1 in- (2.5 cm-) diameter roll and slicing into ½ in (12 mm) pieces. These were then placed in petit-four cases, turned over and pressed down onto the up-turned base of a cut-glass wine glass to make the pattern on the top. The sides were then rolled against the sides of the petit-four case to give a ridged effect. Set aside for a couple of days in a cool place to harden.

Elizabethan Posy

....

Red velvet 30 in x 7 in (75 cm x 17.5 cm)
Circle of card 5 in (12.5 cm) in diameter
Half of a 4 in- (10 cm-) diameter florist's
dry foam cylinder
Latex adhesive
Wadding 1 in (2.5 cm) x 18 in x 12 in
(45 cm x 30 cm)
Lace 30 in x 4 in (75 cm x 17.5 cm)
Dried red roses
Florist's reel wire
Bleached quaking grass (Briza maxima)

With right sides facing, stitch the short edges of
the velvet together. Gather stitch one long edge,
draw up tightly and fasten securely with stitches to
form a 'bag'. Position the card circle in the center

bottom of the 'bag' and secure the foam onto the center of the card. Roll the wadding into a sausage shape and arrange around the foam to pad out the 'bag'. Turn under 1 in (2.5 cm) along the top edge of the 'bag' and gather stitch. Pull up the gathering to the outer edge of the foam and secure with stitches. Stiffen the lace with spray starch and stitch the short edges together. Gather stitch one long edge, draw up and stitch to the top edge of the velvet to complete the 'bag' base. Push stems of dried roses into the foam, longer ones in the center, working down in height, to very short stems around the edge of the foam. Bind small bunches of quaking grass with reel wire, leaving 2 in (5 cm) 'legs' to push into the foam between the roses to complete the arrangement.

Peach Rose Candle-holder

Circle of card 5 in (12.5 cm) in diameter
Quick-drying adhesive or glue gun
Dried eucalyptus leaves
Peach jute scrim ribbon 3 yd (2.75 m) x
3 in (7.5 cm)
Florist's reel wire
Dried peach roses
Gypsophila (Baby's Breath), dyed peach
Candle

Remove a 3 in- (7.5 cm-) diameter circle from the center of the card, to make a ring. Glue overlapping eucalyptus leaves to the ring to cover it. Measure 2 in (5 cm) from the end of the ribbon, bend in half to form a loop and secure with a length of reel wire twisted around the base. Measure 3 in (7.5 cm) along the ribbon from the loop and make a second loop, again securing with reel wire. Repeat a further three times. Glue the ribbon around the ring at each loop base. Glue a single rose behind each loop. Then, glue roses onto the ring between

each loop. Glue the bases of single eucalyptus
leaves around the inside edge of the ring so that
they stand upright, to frame the candle. Cut the
gypsophila into short sprays, apply a small amount
of glue to the bases and intersperse between the
roses and ribbon loops. Push the candle up through
the center of the base.

$\mathcal{S}ensuous\ \mathcal{S}oaps$

½ cup melted solid vegetable oil
½ cup almond oil
¼ cup creamed coconut (culinary)
½ cup cold water
2 tbsp caustic soda (NB This substance is corrosive, therefore precautions need to be taken in its handling. Please follow manufacturer's instructions.)
¼ cup triple-strength rosewater
20 drops rose essential oil
3 drops red food coloring

Before starting work, protect your work surface with several layers of waste paper. Put on an apron or overall and rubber gloves. In the top of a double saucepan, add the melted vegetable oil to the almond oil and creamed coconut. Heat to about 136°F (70°C). Strain the mixture and keep warm. Put the cold water in a deep bowl, carefully add the caustic soda (follow manufacturer's instructions

30

on the container) and mix with a plastic or wooden spoon. The temperature of the mixture will increase as the caustic soda dissolves. Slowly stir in the warmed oil and coconut mixture and beat with a hand or electric whisk for about a minute and a half, when it will thicken. Add the rosewater, essential oil and food coloring, and whisk for a further two minutes. You can now pour the mixture into molds to set. Any suitable plastic container will do, for example thoroughly washed individual mousse or yogurt pots. Alternatively, use the molded plastic linings from boxes of chocolates, choosing those with attractive designs on their bases. You can also use suitable confectionary, chocolate and candle molds. Fill the molds about a quarter full and leave the soap to set for two or three days.

Turn them out onto some folded kitchen paper towels on a plate and leave to dry in a warm cupboard for several weeks to harden before use.

Rose Hand Gel

3 tbsp (2 UK tbsp or 18 ml) glycerine
¼ tsp borax
3 tbsp (2 UK tbsp or 18 ml) cornstarch (cornflour)
1¼ cups (10 fl oz or 300 ml) rosewater
(normal strength)
3 drops rose essential oil
2 drops red food coloring

Place the glycerine in the top of a double boiler and heat gently. Add the borax, then stir in the cornstarch (cornflour) a little at a time and beat into a smooth paste. Continue to heat and add the rosewater a little at a time, stirring continuously, until the mixture becomes thick and creamy. Add the rose oil and food coloring, give the mixture a final beat and pour into a storage jar.

Almond and Rose Oil

There is no better way to condition your skin after bathing than to massage in a little of this oil. You can mix any amount, large or small, provided that you follow the ratio given below, ie 20:1.

2¼ cups (18 fl oz or 500 ml) almond oil
½ tsp rose essential oil

The latter must never be used directly on the skin - it is far too powerful. For our blend, it is worthwhile selecting a top quality essential oil - price usually indicates quality. When mixing, use rubber gloves, shake the mixture well and leave for two days to blend. If using a clear glass jar, do not display in a sunny or hot bathroom. To keep the oil in optimum condition, store in a cool, dark cupboard.

33

Ancient and Modern

Calico
Pressed miniature pink roses, buds and leaves
Pressed forget-me-nots
Stamp tweezers
Latex adhesive
Toothpick
Iron-on protective film laminate
Double-sided adhesive tape
Card
Silk cord

Select a box or book that is the size of the bag you want to make. Cut a piece of calico to fit around the 'mold' allowing an extra 1½ in (3.75 cm) on the width and the same amount as the depth of the mold on the length plus an extra 2 in (5 cm) for folding under the top edge. Therefore, if your mold measures 10 in x 7 in x 3 in (25 cm x 17.5 cm x 7.5 cm), you will need a piece of calico measuring 21 ½ in x 15 in (53.75 cm x 37.5 cm). With pins, mark out what will be the front side of the bag in the center of the length of calico, so that the seam will fall down the center back. Pick up each flower in turn with tweezers and apply a small dot of adhesive to the back with a toothpick. Position in the center of the front panel and make an attractive

2 in (5 cm) along the top edge of the calico and crease. Wrap the fabric around the mold and seal the side edges together at the back with double-sided adhesive tape, folding in the rough edges. Cut a piece of card to match the width and depth of the mold for the bag's base and position inside the calico at the bottom of the mold. Crease and fold the sides of the bag in to the card base, securing to the base with adhesive tape, to leave two flaps, front and back. Crease and fold in the back flap. Seal to the base with adhesive tape. Repeat with the front flap. Remove the bag from the mold. Crease the side edges and fold in the sides. Fold up the base to meet the back and place between heavy books to impress folds. Secure a 1 ½ in- (3.75 cm-) wide strip of strengthening card to the inside front and back sides along the top edge of the bag, under the flap of fabric, before punching two holes in each. Thread a length of cord through each pair of holes and knot the ends inside the bag to secure.

shape by adding a few leaves and buds. Use the remaining buds to make a border around the outer edge of the panel. Remove the pins and, following the manufacturer's instructions, apply the plastic laminate to the right side of the calico. Fold under

Cutting and Preparing

Pick garden roses either early in the morning or in the evening, when the flowers' cells are full of moisture. Cut stems cleanly on a slant with a very sharp pair of pruning shears, scissors or knife.

Remove thorns, and all stems that will be submerged below the water level should be cleaned of leaves or branches, which otherwise will rot and cloud the water. The bottom of the stems should be thoroughly split or crushed to allow water to travel up them. Leave roses standing in a bucket of tepid water in a cool place to have a long drink before arranging.

Left: *Even roses that have been bred almost thornless need to be cleaned along their stems. Remove leaves and thorns by scraping with the edge of a scissor blade.*

36

Sweet Simplicity

What could be simpler and yet more stunning than an elegant, tall-stemmed glass of beautiful roses. The secret of success here is to densely pack the glass with a selection of blooms in varying shades of the same color: in this case, from palest pink to deep crimson touches in the form of small buds. For additional interest, we have included some old-fashioned roses - notably Ferdinand Pichard, with striped crimson petals. This variety also offers a powerful scent.

Right: *A romantic dinner on a warm summer evening* en plein air, *crowned by a glass of intoxicating roses, adding scent, color and texture to a beautiful setting.*

Hearts' Desire

St Valentine's Day is a marvelous opportunity to be hopelessly romantic and create a decoration or gift in true Victorian style. A heart-shaped basket provides the perfect container for a dramatic collection of red roses, the very embodiment of love. The basket is first filled with damp floral foam. The stems of the scarlet rosebuds are cut very short and then packed into the center of the basket following the heart shape of the container. This solid mass of color and texture ensures maximum impact. An edging of lilac serves to soften both the hard edges of the basket as well as the solid outline of the heart arrangement of roses.

Heart-shaped fresh rose decorations created solely from roses of a single color can generate very different moods, depending on the color chosen: vermillion for powerful drama; shaded pinks for a soft and pretty effect.

Hybrid Tea

To recreate this sumptuous centerpiece, fill a large, round rose bowl or attractive china tureen, as in our picture, with damp foam or crumpled wire, if you prefer. Add umbellifer flowers all over, combined with touches of variegated foliage. Fill in with roses, working all round the bowl to achieve a domed outline.

Florist roses can never rival the majesty of garden varieties. When the latter are in season, be sure to enjoy them to the full by displaying them simply and prominently. Pick bunches of blooms and arrange loosely in a shallow bowl.

Right: *A gloriously informal table decoration of old-fashioned garden roses, fit for the fanciest of celebrations.*

Red and White Roses

Read in these roses the sad story
Of my hard fate, and your own glory.
In the white you may discover
The paleness of a fainting lover;
In the red the flames still feeding
On my heart, with fresh wounds bleeding.
The white will tell you how I languish,
And the red express my anguish;
The white my innocence displaying,
The red my martyrdom betraying.
The frowns that on your brow resided,
Have those roses thus divided.
Oh! let your smiles but clear the weather,
And then they both shall grow together.

Thomas Carew (?1595-?1640)

Right: Madonna of the Rose Bower, *Martin Schöngauer (1440-91), St Martin, Colmar, Germany.*

The Sick Rose

O Rose, thou art sick!
The invisible worm
That flies in the night,
In the howling storm,
Has found out thy bed
Of crimson joy:
And his dark secret love
Does thy life destroy.

William Blake (1757-1832)

Below: Miniature of a Young Man Against a Rose Tree, *Nicholas Hilliard (1547-1619), Victoria and Albert Museum, London, England.*

Left: Marie Antoinette, Queen of France, Wife of Louis XVI, Holding a Rose, *Élisabeth Vigée Lebrun (1755-1842), Château Versailles, France.*

To a Friend

As late I rambled in the happy fields,
What time the skylark shakes the tremulous dew
From his lush clover covert; - when anew
Adventurous knights take up their dinted shields:
I saw the sweetest flower wild nature yields,
A fresh-blown musk rose; 'twas the first that threw
Its sweets upon the summer: graceful it grew
As is the wand that queen Titania wields.
And, as I feasted on its fragrancy:
I thought the garden rose it far excell'd:
But when, O Wells! thy roses came to me
My sense with their deliciousness was spell'd:
Soft voices had they, that with tender plea
Whisper'd of peace, and truth, and friendliness
unquell'd.

To a Friend Who Sent Me Some Roses,
John Keats (1795-1821)

Left: Pink and White Roses, *Henri Fantin-Latour
(1836-1904), private collection.*

47

Rosa Gallica Regalis.　　　*Rosier Grandeur Royale.*

Of Poetry and Painting

June of the iris and the rose.
The rose not English as we fondly think.
Anacreon and Bion sang the rose;
And Rhodes the isle whose very name means rose
Struck roses on her coins;
Pliny made lists and Roman libertines
Made wreaths to wear among the flutes and wines;
The young Crusaders found the Syrian rose
Springing from Saracenic quoins,
And China opened her shut gate
To let her roses through, and Persian Shrines
Of poetry and painting gave the rose.

Vita Sackville-West (1892-1962)

Left: Rosa Gallica Regalis, *Pierre-Joseph Redouté*
(1759-1840), private collection.

49

Through the Looking Glass

"There's one other flower in the garden that can move about like you," said the Rose, "I wonder how you do it -" ("You're always wondering", said the Tiger-lily), "but she's more bushy than you are."

"Is she like me?" Alice asked eagerly, for the thought crossed her mind, "There's another little girl in the garden, somewhere!"

"Well, she has the same awkward shape as you", the Rose said, "but she's redder - and her petals are shorter, I think."

"Her petals are done up close, almost like a dahlia," the Tiger-lily interrupted: "not tumbled about anyhow, like yours."

"But that's not your fault", the Rose added, "you're beginning to fade you know - and then one can't help one's petals getting a little untidy."

Lewis Carroll (1832-1898)

Right: Gabrielle with Rose, *Pierre-Auguste Renoir (1841-1919), Louvre, Paris, France.*

The Development of Roses

Roses are native to the Northern Hemisphere, mainly China but also Europe and North America. Some botanists claim there are over 3,000 distinct species of roses, but the number of good ones is no more than 150 and only a relatively few of these have contributed to the wealth of roses grown today. Roses are very promiscuous, freely hybridizing with each other as well as allowing man to manipulate them to create an even wider range.

During the Middle Ages, roses were mainly grown in monasteries, often for religious and medicinal uses. Native species hybridized with each other and created natural seedlings of new varieties, while others produced mutations (sports), a process which has occurred for thousands of years. For example, in 1954 the Dog Rose *(Rosa canina)* by chance produced the beautiful 'Abbotswood'.

The Alba, Centifolia, Damask, Gallica and Musk roses were supreme until the introduction into Europe of our Chinese hybrids between 1792

and 1824. These roses helped to create a wider color range, as well as having the ability to flower recurrently throughout summer. Before this, most roses flowered once and were devoid of bloom the rest of the year.

Further roses were introduced and many were cross-bred, resulting in the Noisettes, Bourbons, Tea roses, Hybrid China and Hybrid Perpetuals.

Most recent have been the Hybrid Tea and Floribunda roses. Miniature roses have enabled roses to flourish in small gardens, while walls, fences, arbors and pergolas can be colorfully clothed with climbers and ramblers. There are even roses that act as ground-cover plants.

Left: Rosa gallica
*'Versicolor', or
Rosa Mundi.*

Left: Rosa canina
'Abbotswood'.

53

Bush Roses

For many years these roses were known as Hybrid Teas and Floribundas, but should now be called 'Large-flowered Bush Roses' and 'Cluster-flowered Bush Roses' respectively.

Large-flowered Bush Roses have large, high-centered flowers borne singly or with side-shoots, in flushes from midsummer to late fall.

Cluster-flowered Bush Roses bear flowers in large trusses, with repeat flowering from midsummer to late fall.

Miniature Roses form small bushes, with the flowers, stems and leaves being in proportion. Heights vary from 6-18 in (15-45 cm), depending on the variety and method of propagation. Those raised from cuttings are shorter than grafted or budded plants. They are ideal for planting in window-boxes and other small containers.

Patio Roses are an unofficial classification, with bush-shaped plants growing about 18 in (45 cm) high. Like the miniature types, leaves, stems and flowers must be in proportion to each other.

Above: Many varieties have been developed for growing in containers.

Left: 'Eye Paint' is a vigorous Bush Rose.

Below: *'Peace' is a world-famous Large-flowered Bush Rose.*

Climbers and Ramblers

Few eyes fail to admire a wall, pergola or arbor richly clothed in roses. Their color range is wide, from bright yellow to soft pinks and those with romantically red hues. For the convenience of gardeners, these are usually divided into two types: **ramblers** have long, pliable stems and usually develop large trusses of small flowers, usually in a single flush; **climbers** have stiffer stems and the flowers are usually larger and the trusses smaller. Ramblers need to have all their old wood cut out each year, while climbers have a more permanent framework, making pruning easier.

Above: *'Shot Silk, Climbing'*.

Right: *'Golden Showers' (climber).*
Left: *'Masquerade, Climbing'.*

WHITE

'Aimée Vibert': climber - double, pure-white

'Albéric Barbier': rambler - yellow in bud, opening to white

'Paul's Lemon Pillar': climber - pale yellow, fading to ivory-white

'Sanders' White': ramber - pure white

YELLOWS AND GOLDS

'Golden Showers': climber - golden-yellow, fading to cream

'Maigold': climber - semi-double and bronze-yellow

'Mermaid': climber - sulphur-yellow

'Sutter's Gold, Climbing': climber - gold, flushed with peach

'Emily Gray': rambler - buff yellow fading to pale yellow

PINK

'American Pillar': rambler - bright pink

'Dorothy Perkins': rambler - powdery pink

'Paul's Himalayan Musk': rambler - blush-pink

'Zéphirine Drouhin': climber - deep rose-pink

CRIMSON

'Crimson Shower': rambler - bright crimson

'Crimson Glory, Climbing': climber - dark, rich velvet-crimson

'Etoile de Hollande, Climbing: climber - deep crimson

'Guinée': climber - deep, velvet-crimson

YELLOW, ORANGE AND PINK BLENDS

'François Jiuranville': rambler - deep pink, shaded fawn and yellow

'Masquerade, Climbing': climber - yellow, pink and red

'Meg': climbing - pink, shading to apricot-yellow

'Shot Silk': climbing - cerise-pink, shot with orange-scarlet and shaded with yellow

Scented Roses

Gloriously rich scents epitomize roses, and although some are best known for colorful flowers, many combine these two qualities. The most common redolence of roses is sweetness, but a few have far more exciting bouquets that seldom fail to capture attention. Here are a few of them, grouped under scents and with their colors:

Clothing walls and bowers

APPLE: 'François Juranville' - double and flowing pink; 'Paul Transon' - copper-orange; 'René André' - soft apricot-yellow; 'Silver Moon' - single and creamy-white

CLOVES: 'Blush Noisette' - semi-double and lilac-pink

FRUITY: 'Leander' - double and apricot-yellow

MUSK: 'Paul's Himalayan Musk' - blush-pink

MYRRH: 'Constance Spry' - clear rose-pink; 'Cressida' - apricot-pink

ORANGE: 'The Garland' - creamy-salmon; 'Veilchenblau' - dark magenta

PAEONY: 'Gerbe Rose' - double and soft pink

PRIMROSE: 'Adélaide d'Orleans' - semi-double

Left: 'Vanity', *a Bush Rose with a raspberry scent.*
Background: 'Fritz nobilis' *has a clover-like redolence.*

Shrub roses

APPLE-LIKE AND SWEET
'Nymphenburg' - salmon-pink shaded cerise and orange-yellow

CLOVER
'Fritz Nobis' - fresh pink with darker shading

LEMONY
'Mme Hardy' - white, blush-tinted when in bud

MUSK
'Day Break' - yellow, opening to light yellow; 'Penelope' - creamy-pink

RASPBERRY
'Adam Maserich' - rich pink
'Cerise Bouquet' - cerise-pink
'Honourine de Brabant' - pale pink, striped and spotted crimson and mauve

SWEET PEA
'Vanity' - deep pink

and creamy-pink; 'Débutante' - clear rose-pink; 'Felicite et Perpétue' - double and creamy-white

SWEET PEA: 'Mme Grégoire Staechelin' - coral-pink with crimson overtones

Rose Gardens

DENMARK: Valbyparken, Copenhagen.

FRANCE: Bagatelle, Bois de Boulogne, Paris; La Roseraie de l'Hay les Roses, Paris; Parc de la Tête d'Or, Lyon.

GERMANY: The Rosarium at Sangar Lausen, Leipzig.

GREAT BRITAIN: Cambridge University Botanic Gardens, Cambridge; Castle Howard, Yorkshire; Corsley Mill, Chapmanslade, Wiltshire; Goodnestone Park, Kent (midway between Canterbury and Sandwich); Mottisfont Abbey, near Romsey, Hampshire; Queen Mary's Rose Garden, Regent's Park, London; Royal Botanic Gardens, Kew, London; Royal Horticultural Society's Garden, Wisley, Surrey; Royal National Rose Society's Gardens, St Albans, Hertfordshire; Sissinghurst Castle Garden, near Cranbrook, Kent.

HOLLAND: Westbroekpark, The Hague.

ITALY: Municipal Rose Gardens, Via de Valle Murcia, Rome.

SPAIN: Rosaleda del Parque del Oeste, Madrid.

SWITZERLAND: Parc de la Grange, Geneva.

UNITED STATES OF AMERICA: American Rose Society Gardens, Shreveport, Louisiana; Boerner Botanical Gardens, Hales Corner, Wisconsin; Columbus Park of Roses, Columbus, Ohio; Cranford Memorial Rose Garden, Brooklyn Botanic, Garden, New York; Descano Gardens, Lacanada, California; Exposition Park Rose Garden, Los Angeles, California; Fort Worth Botanic, Fort Worth, Texas; Greenwood Park Rose Garden, Des Moines, Iowa; Hershey Rose Gardens, Harrisburg, Pennsylvania; Huntingdon Library, Art Collection and Botanical Garden, San Marino, California; James P Kelleher Rose Garden, Boston, Massachusetts; Lake Harriett Rose Gardens, Minneapolis, Minnesota; Memphis Botanic Garden, Memphis, Tennessee; Michigan State University Horticulture Gardens, East Lansing, Michigan; Missouri Botanical Gardens, St Louis Missouri; Morcom Amphitheater of Roses, Oakland, California; Norfolk Botanical Gardens, Bicentennial Rose Garden, Norfolk, Virginia; Queens Botanical Garden, Flushing, New York; Samuell-Grand Municipal Rose Garden, Dallas, Texas; The Marion F Rivinus Rose Garden of the Morris Arboretum, Philadelphia, Pennsylvania; Washington Park Rose Garden, Springfield, Illinois; Woodland Park Rose Garden, Seattle, Washington.

Above: *An award-winning rose garden at Jenkyn Place, Hampshire, England.*